The Food-Energy-Water Nexus 2018

Topical Conference at the 2018 AIChE Annual Meeting

Pittsburgh, Pennsylvania, USA
28 October - 2 November 2018

ISBN: 978-1-5108-7634-7

Printed from e-media with permission by:

Curran Associates, Inc.
57 Morehouse Lane
Red Hook, NY 12571

Some format issues inherent in the e-media version may also appear in this print version.

Printed by Curran Associates, Inc. (2019)

For permission requests, please contact AIChE
at the address below.

AIChE
120 Wall Street, FL 23
New York, NY 10005-4020

Phone: (800) 242-4363
Fax: (203) 775-5177

www.aiche.org

Additional copies of this publication are available from:

Curran Associates, Inc.
57 Morehouse Lane
Red Hook, NY 12571 USA
Phone: 845-758-0400
Fax: 845-758-2633
Email: curran@proceedings.com
Web: www.proceedings.com

TABLE OF CONTENTS

Some abstracts in this book refer to images that were not made available by the publisher

539584 Pittsburgh: Urban Agriculture and the Food-Energy-Water Nexus

Abstract:

Cities rely on external supply chains for food, energy and water. Urban areas are growing and the dependence on reliable supply chains for these resources grows. The interdependence between the infrastructure and connections of providing these resources has become increasingly significant. Urban agriculture provides an opportunity to address Nexus challenges. Urban agriculture benefits include increasing food resilience and improving stormwater management. Ancillary benefits can also include the potential to reduce the amounts of materials sent to landfills (through composting), the reduction urban heat-island effect, and reducing wastewater treatment loads â€" all of which come with energy reduction benefits.

Pittsburgh lost a substantial portion of its population during the transition away from its industrial roots as the â€˜Steel City.â€™ A legacy of the population decrease includes empty and abandoned lots (e.g. over 8,000), aging infrastructure, and areas of decline that have created food deserts and challenges associated with the interdependence of food, energy and water. Pittsburgh has implemented multiple initiatives to address this challenge that will realize the benefits from urban agriculture (e.g. a 23 acre urban farm; community driven revitalizations that include energy efficient housing, stormwater infrastructure, rainwater management, and agriculture spaces).

This analysis examines urban agriculture strategic options, the potential benefits, and the challenges. Options include agriculture options (e.g. geoponic, hydroponic, aeroponic systems), water management options (e.g. water reuse), waste management (e.g. biochar as soil enhancer and carbon storage), and energy option (e.g. distributed generation).

538328 Quantifying Virtual Phosphorus Flows in Interstate Food Trade: Implications for Environmental Sustainability

Abstract:

Phosphorus(P) is one of the essential nutrients for agriculture. It is primarily supplied to crops as inorganic phosphorous fertilizer or though manure application. On farm use of phosphorus efficiency is very low, ranging between 10-15%. Rest of the applied P gets bonded in the soil which may subsequently become available for crop uptake. Current agronomic P recommendations for farmers exclude the accumulated P due to unknown timescales of release. Sustainable management of P is an important topic due to the limited availability of inorganic P sources and its downstream potential to cause eutrophication. Additionally, production of phosphorus fertilizers is highly energy intensive with limited potential of waste stream recovery due to techno-economic constraints. Therefore, an accurate understanding of regional P application and efficiency, combined with interregional food trade is necessary to understand the complexity of P cycle. We combine data on current phosphorus application practices with recommended and existing soil P levels to assess sustainability of P management on farm. Additionally, we use P application data with food trade to quantify phosphorus embodied in the inter-regional trade. This work will enable to identify trends and hotspots in US agriculture P application practices for management of food-water nexus.

We develop a food trade and associated virtual phosphorus model for the U.S. by evaluating over 60 crop items covering commodity groups of cereal grains, fruits, nuts, and vegetables. Regional phosphorus mass balances are constructed using datasets from US Department of Agricultureâ€™s Natural Resource Conservation Service, International Plant Nutrition Institute, and Census of Agriculture. The interstate food trade network is formulated from Freight Analysis Framework and translated to virtual phosphorus network by using crop yields and phosphate fertilizer application rates from Census of Agriculture. Preliminary results indicate that phosphorus fertilizer application differ considerably for crop ranging from 2-10 lb P per ton crop for grains to 10-68 lb P per ton of crop for nuts. To estimate whether calculated P application rates are excessive or necessary for specific soil type, we compare it with recommended P levels in each state. In 2012, 28 states had to recommended levels, including states with highly eutrophic waterbodies such as Ohio, Florida, California, Texas, and Maine. With addition of total P fertilizer across crops in the given year, the soil phosphorus for these states increases by a fraction of 2-4 compared to recommended levels. From food trade perspective, we find that 13 billion US tons of food and 5.3 million US tons of embodied P were transferred across the U.S. Specifically, grains dominate food and total embodied P transfers within the U.S., indicating the importance of considering scale of production. The implications of the study findings will be discussed in the context of balancing on-farm conservation policies for simultaneously enriching soil with nutrients and avoiding excess run-offs.

533778 Towards a Two-Level Superstructure Optimization Framework for Land Use Based on Food-Energy-Water Nexus

Abstract:

Land use optimization is the result of competitions between different land types [1]. The main reasons of these competitions are the quantitative constraints of land scales and corresponding Food-Energy-Water (FEW) resources under multiple conflicting objectives [2, 3]. Despite FEW resources play critical important roles for sustaining and improving human life, increasing production demands and sustainable concerns intensify these competitions due to their limitation. How to coordinate these competitions is a complex task for reasonable land allocation based on limited FEW resources while promoting efficiencies and sustainability [4, 5]. A nexus thinking by considering the FEW flow and interactions in multi-objective land use systems has become apparently necessary for efficient use of FEW resources and trade-off decisions [6, 7]. However, challenges arise in representing quantitative FEW-Nexus that supports land-use decisions while encountering conflict objectives, limited data, and multiple scales [8, 9].

To address these challenges, we develop a multi-objective optimization framework by considering total profit, food production, resources use, and environmental penalty as multiple objectives for land use systems. The proposed framework suggests a two-level superstructure optimization method: (1) In the first level, models of all the production units are developed by data-driven modeling and global optimization methods based on limited realistic data. FEW flow among them are quantified and interlinked to construct interval models, which can be represented as interval superstructures. The small-scale MINLP problems can be solved efficiently due to the limited combinations of land units in the interval subsystems; (2) In the second level, multiple interval models with optimal land and FEW allocations are used to construct the extended systematic network and represented as a large-scale superstructure, which can be solved as a MILP problem. A series of FEW indices are provided for decision-makers to analysis nexus in the system, carry out quantitative assessment based on different objectives, and achieve trade-off solutions. The framework is illustrated by a case study on the crop-livestock system, which shows that it can provide multiple land allocation solutions and predicts corresponding yields for land units within the systems under flexible scales. Computational results from interval models indicate that we achieve optimal solutions in the interval subsystems, and valuable production models for yield prediction with high confidence. The performance of these models can be improved by increasing feedback data [10]. The systematic network provides an efficient method for extending the land allocation solutions to multiple land scales. For multiple objectives, the proposed FEW index can be applied to select strategies for optimal land allocation that maximizing food productivity and minimizing water and energy consumptions.

Reference

[1] Memmah, M. M., Lescourret, F., Yao, X., & Lavigne, C. (2015). Metaheuristics for agricultural land use optimization. A review. Agronomy for sustainable development, 35(3), 975-998.

[2] Bergstrom, J. C., Goetz, S. J., & Shortle, J. S. (2013). Land use problems and conflicts: Causes, consequences and solutions. Routledge.

[3] Keairns, D. L., Darton, R. C., & Irabien, A. (2016). The energy-water-food nexus. Annual review of chemical and biomolecular engineering, 7, 239-262.

[4] Miralles-Wilhelm, F. (2016). Development and application of integrative modeling tools in support of food-energy-water nexus planningâ€"a research agenda. Journal of Environmental Studies and Sciences, 6(1), 3-10.

[5] Simpson, G., & Berchner, M. (2017). Water-energy nexus-Measuring integration: towards a water-energy-food nexus index. Water Wheel, 16(1), 22-23.

[6] Garcia, D. J., & You, F. (2016). The water-energy-food nexus and process systems engineering: a new focus. Computers & Chemical Engineering, 91, 49-67.

[7] Ringler, C., Bhaduri, A., & Lawford, R. (2013). The nexus across water, energy, land and food (WELF): potential for improved resource use efficiency?. Current Opinion in Environmental Sustainability, 5(6), 617-624.

[8] McCarl, B. A., Yang, Y., Schwabe, K., Engel, B. A., Mondal, A. H., Ringler, C., & Pistikopoulos, E. N. (2017). Model Use in WEF Nexus Analysis: a Review of Issues. Current Sustainable/Renewable Energy Reports, 4(3), 144-152.

[9] McCarl, B. A., Yang, Y., Srinivasan, R., Pistikopoulos, E. N., & Mohtar, R. H. (2017). Data for WEF Nexus Analysis: a Review of Issues. Current Sustainable/Renewable Energy Reports, 4(3), 137-143.

[10] Nie, Y., Avraamidou, S., Li, J., Xiao, X., & Pistikopoulos, E. N. (2018). Land use modeling and optimization based on food-energy-water nexus: a case study on crop-livestock systems. 13th International Symposium on Process Systems Engineering; Elsevier; Accepted.

537263 Land Availability, Utilization, and Intensification for a Solar Powered Economy

Abstract:

Among all forms of renewable energy, solar energy is the most promising one with sustainable availability and abundance to meet human beingâ€™s energy need. However, solar energy is a diffuse resource and harvesting it requires vast tracts of land. While most studies estimated that land availability is not a major challenge for the feasibility of a total solar economy, a few recent detailed analyses have raised the problem in land availability in some densely populated countries, using actual solar farm production data and considering other factors such as storage requirement and intermittency of solar availability. However, according to their estimation, countries with relatively low population density, such as the United States, have sufficient land for solar farms to meet all energy demand. In this study, we perform a detailed land requirement analysis in each state of the US for a solar powered future and propose land intensification strategies to resolve possible land use challenges.

Land requirement analysis in this study takes into account several issues that are usually ignored. First, efficiencies of major energy conversion steps from current primary energy to end use are different from those when solar energy is deployed. Second, available land is limited to only part of the urban area (rooftops) and miscellaneous land. Farmland, forest and other reserved land such as national and state parks cannot be utilized for solar plant construction. In addition, we prefer local photons for local use for consideration of minimizing transmission loss and energy security.

Our land requirement analysis shows that for many states there is insufficient currently available land and that other land areas will need to be used, which is contrary to previous report. We propose that farmland is the best choice for land use intensification as it minimizes additional environmental impact to existing habitat and propose several photovoltaic systems for this purpose. We then estimate the power output of these systems and estimate the amount of farmland for each state that is needed to meet energy needs. By land intensification, the energy demand of each state in the US can be locally satisfied.

529950 Microbial Community Profile Versus Water Quality in Urban Watersheds

Abstract:

Microbial community profiles can serve as an indicator for unbran water quality. The aim of this study was to gain insight into bacterial diversity and infer water quality based on selected operational taxonomic units (OTUs). Here we report bacterial communities of 12 watersheds sampled at different frequencies from year 2013 to 2016, were characterized by sequencing the V4 to V5 regions of 16S rRNA genes using 454 pyrosequencing and paired-end sequencing platforms. An nMDS plot of all samples revealed that bacterial communities of the same watershed tended to cluster together and a few watersheds shared similar communities. Effects of temporal variation were more pronounced on bacterial communities within watersheds ($P < 0.01$; 2-Factor PERMANOVA) than spatial variation. Cyanobacteria of *Microcystis* and *Cylindrospermopsis* dominated certain watersheds that tended to have algal blooms. Pairwise correlation of gut and sewage-associated OTUs revealed a strong positive correlation among *Enterobacter*, *Klebsiella* and *Escherichia* and *Clostridium sensu stricto 8*, suggesting that presence of these taxa as possible indicators of impairment. A comparison of the abundances of 13 potential antibiotic resistant bacteria, revealed four watersheds with samples that correlated with higher abundances of potential antibiotic resistant bacteria compared to the other watersheds. Thus microbial community determined by high-throughput sequencing in conjunction with qPCR can be used to indicate quality of water in urban watersheds.

528536 Mechanisms Whereby Microbes Promote Intermediate Soil Moisture Content

Abstract:

The function of microbial communities is inextricably with the soil microstructure. At the pore scale, water content regulates hydraulic connectivity and thereby modulates microbesâ€™ access to aqueous and gaseous substrates. Furthermore, soil bacteria near plant roots produce extracellular polymeric substances (EPS), a polymeric substance that helps modulate soil moisture and promote plant growth. EPS regulates soil moisture (i) by serving as hydrogel, swelling during wet conditions and remaining hydrated during dry conditions, (ii) by promoting soil particle aggregates via capillary forces, (iii) by altering soil surface properties through creating water repellent surfaces. EPS-mediated moisture retention behavior in soil is complicated but can be understood by systematically controlling the microscale environment. Here, we show the effect of physical microstructure on EPS-mediated moisture retention. We employ physically complex emulated soil micromodels as well as micromodels with uniform macropore or micropore geometries. EPS was collected from stationary-phase *Sinorhizobium meliloti* cultures and suspended at different concentrations in growth media salts or in artificial groundwater. Experimental results showed that the 0.25× EPS solution dried eight times slower than deionized water, and the 1× EPS solution dried 16 times slower than deionized water in identical soil micromodels. Also, the residual saturation of 0.25× and 1× EPS solutions were 6 ± 1% and 35 ± 21%, compared with a residual saturation of 0 for deionized water. We observed no effect on rate or extent of moisture loss for the micromodels with a uniform macropore geometry. Studies with uniform micropore geometry confirmed that higher EPS concentrations were associated with a reduce rate and increased variance of moisture loss at the pore scale. In this study, we provide a systematic method to evaluate the performance of EPS solutions in modulating soil moisture, and show how EPS concentration decreases the rate of water loss and also amplifies the variability of moisture content at the pore scale. Longer term, we anticipate that the impact of this work will be to aid in the development of sustainable agriculture biotechnology for increased food production using less water.

528379 Nitrogen Efficient Fertilizer Materials

Abstract:

Nitrogen cycle management is a critical challenge in modern engineering. NH_3 synthesis from N_2 consumes several percent of total world energy while only a fraction of the synthesized fertilizer nitrogen, such as urea, is absorbed by the plants. Majority of nitrogen initially confined in urea is decomposed and released in the environment as gaseous NH_3 or soil and water mobile nitrate, NO_3^-, resulting in low sustainability nitrogen fixation process. Furthermore, for a complete plant nutrient all of the nutrients (N,P,K,Ca,Mg,S) need to be supplied in one solid granulated or liquid mixture. Aqueous liquid fertilizers are expensive to transport since their dissolved nutrient content is governed by the solubility limits.

Molecular crystals of urea with inorganic acids (H_3PO_4 and H_2SO_4) as well as solid salts ($CaSO_4$, $MgSO_4$, $Ca(H_2PO_4)$ and $Mg(H_2PO_4)$) present a convenient form of high nutrient content fertilizers that can be inexpensively transported to the remote locations and applied directly or via fertigation. Importantly, these urea inorganic acid adducts have been shown to result in a much more efficient nitrogen uptake by plants combined with significantly lower nitrogen losses. The work will be presented that will explore design of these fertilizers that contribute to the enhanced sustainability of the chemical processing. In particular, mechanochemical synthesis methods that avoid use of liquid solvents that would otherwise necessitate evaporation or result in high aqueous solution transportation costs will be presented. Since dissolution of the solid urea adducts for fertigation requires in depth knowledge of the corresponding phase diagrams, these data will also be presented. Finally, molecular structure effects of the urea adducts on the improved nitrogen cycle management will be discussed as well as the synthesis of industrial and anthropogenic waste derived nitrogen efficient molecular fertilizers.

541778 Food-Energy-Water Nexus Systems Engineering

Abstract:

Due to population growth, economic development, urbanization, climate change, and natural resource degradation, the global demand for food, energy and clean/freshwater is rapidly increasing [1]. As demand grows the stresses and interdependences between these three resource systems, which are commonly referred to as Food-Energy-Water Nexus (FEW-N), are becoming more apparent [2,3]. Decision making can be very challenging due to multiple dimensions of the biophysical water, energy and food systems, and the players and stakeholders connected with them [4]. A holistic systems engineering approach is thus clearly needed to navigate the multi-faceted FEW-N space, identify opportunities for synergistic benefits and systematically explore interactions and trade-offs [5].

In this work, we present the foundations of a systems engineering framework and quantitative decision-making tools for the analysis and trade-off optimization of stressed interconnected resource FEW-N networks. The framework combines data analytics and mixed-integer modeling and optimization methods establishing the interdependencies and potentially competing interests amongst the food, energy and water elements in the system, along with policy, sustainability and feedback from the various stakeholders. A multi-objective optimization strategy is followed for the analysis of the trade-offs empowered by the introduction of composite FEW-N metrics as means to facilitate decision making and compare alternative process and technological options. The versatility, potential and applicability of the proposed framework is demonstrated by three nexus case studies, (i) a complex power generating system in the South-Central Texas region [6], (ii) a dairy production and processing plant [7], and (iii) a land use allocation problem on a regional farming system in China [8].

References

[1] D'Odorico, P., Davis, K. F., Rosa, L., Carr, J. A., Chiarelli, D., Dell'Angelo, J., ... & Rulli, M. C. (2018). The Global Foodâ€Energyâ€Water Nexus. Reviews of Geophysics.

[2] McCarl, B. A., Yang, Y., Schwabe, K., Engel, B. A., Mondal, A. H., Ringler, C., & Pistikopoulos, E. N. (2017). Model Use in WEF Nexus Analysis: a Review of Issues. *Current Sustainable/Renewable Energy Reports*, *4*(3), 144-152.

[3] McCarl, B. A., Yang, Y., Srinivasan, R., Pistikopoulos, E. N., & Mohtar, R. H. (2017). Data for WEF Nexus Analysis: a Review of Issues. *Current Sustainable/Renewable Energy Reports*, *4*(3), 137-143.

[4] Daher, B., Mohtar, R.H., Pistikopoulos, E.N., Portney, K.E., Kaiser, R. and Saad, W. (2018) Developing Socio-Techno-Economic-Political (STEP) Solutions for Addressing Resource Nexus Hotspots, Sustainability 2018, 10, 512; doi:10.3390/su10020512

[5] Garcia, D. J., & You, F. (2016). The water-energy-food nexus and process systems engineering: a new focus. *Computers & Chemical Engineering*, 91, 49-67.

[6] Diangelakis, N. A., & Pistikopoulos, E. N. (2017). A multi-scale energy systems engineering approach to residential combined heat and power systems. *Computers & Chemical Engineering*, 102, 128-138.

[7] Avraamidou, S., Milhorn, A., Sarwar, O., & Pistikopoulos, E. N. (2018). Towards a quantitative Food-Energy-Water Nexus metric to facilitate decision making in process systems: A case study on a dairy production plant. *28th European Symposium on Computer-Aided Process Engineering*: Accepted.

[8] Nie, Y., Avraamidou, S., Li, J., Xiao, X., & Pistikopoulos, E. N. (2018). Land use modeling and optimization based on food-energy-water nexus: a case study on crop-livestock systems. *13th International Symposium on Process Systems Engineering*: Accepted.

Renewable Carbons from Food Waste for Separation and Catalysis Technologies

This study explores the production of renewable and sustainable carbons from waste biomass resources, for adsorption and catalytic applications. Specifically, the research will explore the production of carbons derived from **food waste** and their utilization as sorbents of aromatic contaminants in **water** and catalyst supports for the production of clean **energy and fuels**.

The idea of using renewable and sustainable resources to replace the carbon derived from fossil fuels is not new. Researchers have been working on the utilization of various biomass resources, such as rice husk, coconut shells and others for years.[1–5] However, the utilization of food waste as a carbon resource has not been extensively explored. Food waste could be the ultimate resource for carbon production, since it is produced in excessive volumes and it represents one of the most significant environmental problems, which is urgently looking for a solution. The critical question is whether the activated carbons derived by food waste have the potential to be used for purification and/or catalysis purposes. Food waste is heterogeneous in nature and contains various types of minerals, which may provide advantages or disadvantages, depending on the application.

In our lab, we receive the food waste from the Dining Halls of the University of Connecticut and after pretreatment, that includes washing, grinding and sieving, we perform pyrolysis and activation to turn the waste into carbons. The produced carbons are then characterized and they are tested as sorbents and catalyst supports. Our goal is to understand how the conditions of pyrolysis and activation affect the properties of the carbons and, consequently, their effectiveness as sorbents and catalyst supports. During the pyrolysis reaction, temperature and residence time are the most important parameters that affect the yield and the properties of the produced biochar. Activation of the biochar is performed either using steam, physical activation, or using chemicals (e.g. H_3PO_4), chemical activation. During the former method of activation, temperature and residence time are, again, among the most important parameters that affect the surface area, the carbon content and the porosity of the produced carbons. During the latter method, the impregnation ratio between the chemical and the biochar is the most critical.

During this talk we will discuss the properties of the activated carbons produced by food waste and how these properties are affected by the reaction conditions and the method of activation. Surface area, porosity, carbon content, surface functional groups, morphologies are the properties of interest which will be discussed. Then, we will discuss the potential of using the activated carbons produced by food waste, as sorbents of aromatic hydrocarbons in water and catalyst supports for the hydrodeoxygenation of bio-oil.

References

(1) Tripathi, M.; Sahu, J. N.; Ganesan, P. Effect of Process Parameters on Production of Biochar from Biomass Waste through Pyrolysis: A Review. *Renew. Sustain. Energy Rev.* **2016**, *55*, 467–481.

(2) Kwapinski, W.; Byrne, C. M. P.; Kryachko, E.; Wolfram, P.; Adley, C.; Leahy, J. J.; Novotny, E. H.; Hayes, M. H. B. Biochar from Biomass and Waste. *Waste and Biomass Valorization* **2010**, *1* (2), 177–189.

(3) Ahmad, M.; Rajapaksha, A. U.; Lim, J. E.; Zhang, M.; Bolan, N.; Mohan, D.; Vithanage, M.; Lee, S. S.; Ok, Y. S. Biochar as a Sorbent for Contaminant Management in Soil and Water: A Review. *Chemosphere* **2014**, *99* (November 2013), 19–23.

(4) Tan, Z.; Lin, C. S. K.; Ji, X.; Rainey, T. J. Returning Biochar to Fields: A Review. *Appl. Soil Ecol.* **2017**, *116* (September 2016), 1–11.

(5) Azargohar, R.; Dalai, A. K. Biochar as a Precursor of Activated Carbon. *Appl. Biochem. Biotechnol.* **2006**, *131* (1–3), 762–773.

541779 A Computational Framework for Sustainable Waste Management and Simultaneous Recovery of Nutrients and Energy

Abstract:

To protect human health and the environment, communities should have sustainable infrastructures to manage waste. Critically, livestock waste may cause some air quality degradation from ammonia and methane emissions, soil quality detriment due to in-excess nutrients and acidification, and water pollution issues resulting from nutrient and pathogens runoff to water bodies, which leads to eutrophication, hypoxia, and harmful algal blooms (HABs).

In addition, the recovery of carbon from organic waste offers an alternative for replacing fossil fuels with renewable sources for energy production. Despite the significant environmental benefits by performing pollution management of these organic materials, the deployment of technologies for obtaining value-added products and renewable energy from livestock waste is not a current practice due to the high investment costs and low product market values.

This contribution describes a multi-scale approach for the development of a supply chain framework to identify and evaluate techniques and logistics for the management of solid waste generated from primary sources of nutrient pollution. This effort will inform stakeholdersâ€™ waste management options while assessing potential economic and environmental benefits. Other potential benefits include the reduction of nutrient pollution and HABs, recovery of valuable products, and adoption of nutrient-reducing technologies in different locations (e.g., community, state, regional, etc.). In addition, the proposed framework captures techno-economic and logistical issues and can accommodate diverse types of material management incentives obtained at federal and state levels.

540773 2018 Outlook for Energy: A View to 2040

Abstract:

The Outlook for Energy is ExxonMobil's global view of energy demand and supply through 2040. It also highlights the dual challenge of providing affordable energy to support prosperity while reducing environmental impacts – including the risks of climate change. We use the Outlook to help inform our long-term business strategies and investment plans.
The Outlook anticipates significant changes through 2040 across the world to boost living standards, reshape the use of energy, broaden access to abundant energy supplies, and accelerate decarbonization of the world's energy system to address the risks of climate change.

- Global energy needs will rise about 25 percent over the period to 2040
- Electricity demand will nearly double in non-OECD countries
- Worldwide electricity from solar and wind will increase about 400 percent
- Natural gas will expand its role, led by growth for electricity generation
- Growth in oil demand will be driven by commercial transportation and chemical needs
- Efficiency gains and growing use of less-carbon intensive energy sources will contribute to a nearly 45 percent decline in the carbon intensity of global GDP

Energy is fundamental to modern life, and as the world's population approaches 9 billion people in 2040, we are challenged to improve living standards everywhere. Understanding the factors that drive the world's energy needs – and likely choices to meet those needs – is the mission of the Outlook. By sharing the Outlook with the public, we hope to broaden that understanding. Energy matters to everyone, and we all play a role in shaping its future.

541543 Energy Decarbonisation Scenarios

Abstract:

With the energy sector being associated with 2/3 of the annual greenhouse gas emissions, decarbonizing the energy is a key step to reduce their level. The National Determined Contributions (NDC's) announced so far by the participating countries do not meet the objectives of the Paris-climate agreement, as far as reaching a peak in emissions in the near future or decreasing to a level that is compatible with the stated long-term targets.

Using different scenarios for energy demand and supply, options for energy sector decarbonisation are evaluated. Meeting the 2-degree target requires a major transformation in the power sector, efficiency improvements – specially in the buildings and in the industry, and a revolution in transport towards fuel and modal shifts. With a significant increase in renewable energy, system integration and optimization will also be key in reducing the cost of the energy transition. Meeting more ambitious climate goals ("well below 2 degrees") will push further the technological innovation and implementation requirements throughout the entire energy value chain. A scenario aiming at zero net-emissions by 2060 is presented, along with a comparison to the 2-degree case.

542625 Fundamental Research Needs to Advance Energy Technologies

Abstract:

The U. S. Department of Energy (DOE), Office of Science, Office of Basic Energy Sciences (BES) supports fundamental research in chemical and materials sciences to provide the foundations for new energy technologies and to support DOE missions in energy, environment, and national security. This presentation will discuss opportunities for fundamental research to impact DOE's energy mission "to catalyze the timely, material, and efficient transformation of the nation's energy system and secure U.S. leadership in energy technologies" with a focus on the way we generate, store and use energy nationally. I will provide an overview of BES strategic planning over the past decade that identified priority research directions for advancing energy applications, highlight key scientific advances in these areas, and discuss some future opportunities for modern science, particularly chemical sciences, to accelerate the transformation of the U. S. energy portfolio.

542591 Electrochemical Conversion of Ammonia and Nitrogen for Sustainable Food-Energy-Water

Abstract:

Electrochemical Conversion of Ammonia and Nitrogen for Sustainable Food-Energy-Water

Gerardine G. Botte

Center for Electrochemical Engineering Research

Department of Chemical and Biomolecular Engineering

Russ College of Engineering and Technology

Ohio University, Athens

OH 45701

The production of reactive nitrogen fertilizer via the Haber-Bosch (HB) process, combined with industrialization and intensification of agriculture, has created an abundant global food supply to support population growth, but has doubled the amount of reactive nitrogen compared to pre-industrial levels.[1] With those perspectives, ammonia can be seen as an important chemical of use or as an environmental challenge. Ammonia emissions into air (ambient ammonia) and water represent an environmental challenge. Ambient ammonia not only contributes to inorganic PM2.5 (particulate matter with an aerodynamic diameter of less than 2.5 Î¼ m) directly but also plays an important role in secondary organic aerosol formation by interacting with gaseous phase organic acids and forming condensable salts. In addition, ammonia emissions in water are associated with environmental problems such as algae bloom. Various industries and other operations are considered ammonia emitters, but the ones related to sustainable food production such as livestock management and fertilizer application are among the major contributors accounting for 60-75% of the ammonia emissions.[2-3]

On the other hand, ammonia produced commercially via the HB process is usually known as one of the greatest inventions of the 20th Century. The HB process made it possible for the first time to produce synthetic fertilizers and produce sufficient food for the Earthâ€™s growing population. However, the HB process requires high temperatures and pressures and as a result it can only be done in large plants which consume large amounts of power and generate large quantities of CO_2 from the fuels used to power the process, such as natural gas. It is estimated that the HB process accounts for nearly 1% of the entire global power consumption. Due to the sheer scale of the HB process, distribution from the point of production to the point of use becomes an additional carbon and energy burden. Alternative methods of manufacturing ammonia that enable small scale, distributed generation at the point of use using renewable energy and sustainable feedstocks could have a great impact on global CO_2 emissions.

A closed loop approach that includes producing fertilizer at the point of use, applying nitrogen fertilizer as close to the time that crops needs it, and recovering and recycling nitrogen fertilizer can contribute to address the challenges related to ammonia emissions, while decreasing water and energy consumption. Electrochemical conversion of nitrogen and ammonia presents an excellent platform to achieve this closed loop approach. Dr. Botte and members of her research group at the Center for Electrochemical Engineering Research (CEER) are working on the electrochemical conversion[4]/synthesis of ammonia[5] in alkaline media. In this talk, recent advances in the platform process for ammonia and nitrogen removal/recycling and ammonia synthesis will be discussed as well as their application for sustainable food, energy, and water.

References:

1. Grand Challenges - Manage the Nitrogen Cycle [WWW Document], n.d. URL http://www.engineeringchallenges.org/9132.aspx (accessed 5.28.18).

2. G. E. Mansell, "An Improved Ammonia Inventory for the WRAP Domain (Final Report)", ENVIRON International Corporation, Report No. 415.899.0700, March 7 (2005).

3. FAO. Global estimates of gaseous emissions of NH_3, NO, and N_2O from agricultural land, International Fertilizer, Industry Association, Rome. 2001.

4. Botte, G. G., Cooper, M. and Vitse, F., (2009) Electro-catalysts for the oxidation of ammonia in alkaline Media, U.S. Patent No. 7,485,211.

5. B. L. Sheets, G. G. Botte, Electrochemical nitrogen reduction to ammonia under mild conditions enabled by a polymer gel electrolyte, Chemical Communications 54 (2018), 4250-4253.

542589 Encapsulation and Nanoparticle Formation for "Non-Standard" Applications

Abstract:

Encapsulation and nanoparticle formation for "non-standard" applications

Prof. Robert K. Prud'homme[1], Prof. Rodney D. Priestley[1], Leslie Shor[2] ,Douglas Scott[1], Jie Feng[1]

[1] Princeton University, Department of Chemical and Biological Engineering, Princeton University, Princeton NJ 08544

[2] University of Connecticut, Department of Chemical and Biomolecular Engineering, Storrs, CT 06269-3222

Presenting author: Robert K. Prud'homme, prudhomm@princeton.edu

Traditionally, encapsulation and nanoparticles have been associated with high value applications such as drug delivery in medicine. The two main reasons have been the high cost of nanoparticle assembly processes for complex nanoparticles that encapsulate and release actives in a controlled way, and the fact that encapsulation technologies for larger scale macro-particles do not translate to nano-scale particles. Recent advances in low-cost nanoparticle assembly present the opportunity to open up new applications. First, we present an overview of encapsulation and release technologies. Then we provide a review of the rapid-precipitation technologies we have developed that rely on the control of the fluid dynamics, diffusion-limited aggregation, and steric stabilization of nanoparticles. The processes are scaleable, low cost, and produce nanoparticles from 50-400 nm. Several novel applications will be presented: (1) nanoparticle delivery of agricultural actives s to plant roots by protist active transport, (2) triggered indicator inks for sterilization processes, (3) "green chemistry" encapsulation of personal care compounds using purely aqueous processing.

522027 Energy-Water Nexus Study for a Mushroom Farming Initiative in Nigeria

Abstract:

A water treatment and recycle system was designed for a mushroom farming initiative in Kaduna, Nigeria. The energy requirements of the proposed system were studied to determine the optimal system to minimize energy use for water treatment. Other considerations were ease of use and capital and operating costs.

The concept behind the cooperative is for 200 women farmers pool their resources to pay for water and energy costs. Electricity will be provided through solar panels. We were consulted to design a water treatment and recycle system for a demonstration of the concept. The concept will likely be replicated in remote areas without an abundance of water, so water recycle is essential. Reliable electricity will also likely be an issue so solar panels will be used.

The first step was to collect samples of water from the potential water source: the Kaduna River. Preliminary tests showed that the water was relatively clean but contained bacteria and some hardness. A treatment and recycle system was designed appropriate for the impurities in the water. Due to limited energy provided by the solar panels, energy requirements were studied for all appropriate treatments to minimize energy use of the final system. The final design and final energy use of the water system will be presented.

523808 Sustainable Optimal Strategic Planning for Shale Water Management

Abstract:

The development of new additives in the shale gas industry, which tolerate the use of high Total Dissolved Solids (TDS) base fluid, has allowed reusing the wastewater in the drilling of subsequent wells (U.S. Environmental Protection Agency, 2016). This practice is currently the most popular and cost-effective option for shale gas water management. Although it minimizes freshwater consumption, producers should take into consideration possible long-term issues and challenges. For instance, the TDS concentration will increase significantly, around 2-3 times, which can represent a cost barrier to reuse the water for fracturing operations. Besides, as the number of drilled wells decrease this practice becomes less attractive. Specifically, the volume of fracturing fluid required to fracture new wells may be less than the volume of water generated by producing wells in the area.

Currently, several works have been reported on the optimization of shale gas water management (Gao and You, 2015; Guerra et al., 2016; Yang et al., 2015). However, most of the works published in the literature consider that the water blending ratio is restricted or the return to pad operations are not allowed. Drouven and Grossmann (2017) assume that the water-blending ratio is unrestricted, they over-estimate the friction reducers expenses. The MILP model that they proposed does not account the salt concentration of impaired water. Moreover, they only distinguish between two types of water: impaired water and freshwater. Hence, the model cannot handle any water management option when drilling operation decrease.

In this work, we propose a mixed-integer non-linear programming (MINLP) model considering the TDS concentration of flowback and impaired water as a function of time and for different water treatment solutions. We estimate the friction reducers expenses as a function of TDS concentration to determine if the level of TDS in impaired water is an impediment to reusing it in fracturing operations. Moreover, the model distinguishes between four types of water: impaired water, flowback water, desalinated water and freshwater. The objective is to maximize the "sustainability profit" in order to obtain a compromise solution between economic, environmental and social aspects. Only a single objective function is necessary since all the indicators are expressed in monetary terms (Zore et al., 2017).

The water management system comprises wellpads, shale gas wells in each wellpad, centralized water treatment technologies (CWT), freshwater sources, fracturing crews, and disposals wells.

After hydraulic fracturing, a portion of water called flowback water returns to the wellhead. The flowback water is stored in fracturing tanks (FT) onsite before basic treatment (pre-treatment) in mobile units or transport to CWT facility or Class II disposal. Pre-treatment includes technologies to remove suspended solids, oil and grease, and bacteria, certain ions that can cause the scale to form on equipment and interfere with fracturing chemical additives. After pre-treatment, the water can be used as a fracturing fluid in the same wellpad or in the neighboring wellpad, or can be desalinated in the onsite TDS removal technologies. The flowback water reused for fracturing operations is called impaired water and is stored in impaired water tanks (IWT). Several desalination technologies can be selected such as multi-stage membrane distillation (MMD), multi-effect evaporation with mechanical vapor recompression (MEE-MVR) or forward-reverse osmosis (FO-RO) hybrid. We consider that the outflow brine salinity in the onsite treatment is close to salt saturation conditions to achieve ZLD operation. Cost and salinity levels restrict the type of desalination unit that can be used for TDS removal. The onsite desalinated water can also be used as a fracturing fluid in the same wellpad, transported to the next wellpad or discharge for other usages. The flowback water can also be transported and treated in CWT. Desalinated water from CWT can select the same routes as the desalinated water in onsite technologies. Freshwater is obtained from an uninterruptible fresh water source. Desalinated water and freshwater are stored in freshwater tanks (FWT) and/or water impoundment. Transportation of freshwater, impaired water, flowback water and desalinated water can only be through trucks. Storage tanks and mobile treatment are assumed to be leased.

The problem is to determine: (1) wellpad fracturing start date (fracturing schedule), (2) number of tanks leased at each time period, (3) number of trucks needed in each time period, (4) flowback destination, reuse (impaired water) , treatment (onsite or offsite) or disposal, and (5) quality of water used to fracture each well.

The optimization problem is formulated as an MINLP model that includes: assignment constraints, material balance in storage tanks, mixer and splitters, logic constraints and an objective function. Salt material balances are modeled using total flows and salt composition as variables (bilinear terms). An advantage of using this representation is that the variables involved in the bilinear terms are well bounded, allowing us to define tight under and over estimators.

Them objective function includes the economic profit, the eco-profit and the social-profit.

- *Economic profit* includes revenues from natural gas minus the sum of the following expenses: wastewater disposal cost, storage tank cost, freshwater cost, friction reducer cost, wastewater and freshwater transport cost and onsite and offsite treatment cost.
- *Eco-profit* distinguishes between eco-benefit (raw material and products that unburden the environment) and eco-cost (raw material al products that burden the environment). Both terms are calculated by using eco-cost coefficients (Delft University of Technology, 2017). In our problem, impaired water and desalinated water used to fracture a neighboring well exhibit unburdening effect on the environment. However, natural gas, freshwater withdrawal, disposal and transportation burden the environment.
- *Social profit* includes social security contributions paid for the employed people to fracture a well, plus the social transfer by hiring people, minus social cost. We only take into account the numbers of jobs on a fracturing crew and the time that they are working to fracture a specific well. Once the well is completed, the number of jobs generated by truck drivers or maintenance team are not considered.

The proposed model is applied to a case study in Marcellus Play with 3 wellpads and 20 wells, three years discretized at one week per time period, four interruptible sources of fresh water, three class II disposal wells, two CWT plants and one fracturing crew. The MINLP problem consists of 2808 binary variables, 14977 continues variables and 12010 constraints.

The solution of the model shows that the capabilities to reuse impaired water to fracture other wells is the best economic and environmental practice for shale gas water management. The model, which can be effectively solved with the proposed decomposition technique, reveals that the level of TDS in impaired water is not an obstacle to reusing it for fracturing purposes. Also, it has been shown that onsite desalination treatment can be cost-effective for operators once no more wells to fracture are available.

References:

Delft University of Technology, 2017. The Model of the Eco-costs/Value Ratio (EVR). URL http://www.ecocostsvalue.com (accessed 12.1.17).

Drouven, M.G., Grossmann, I.E., 2017. Optimization models for impaired water management in active shale gas development areas. J. Pet. Sci. Eng. 156, 983–995.
Gao, J., You, F., 2015. Shale Gas Supply Chain Design and Operations toward Better Economic and Life Cycle Environmental Performance: MINLP Model and Global Optimization Algorithm. ACS Sustain. Chem. Eng. 3, 1282–1291.
Guerra, O.J., Calderón, A.J., Papageorgiou, L.G., Siirola, J.J., Reklaitis, G. V., 2016. An optimization framework for the integration of water management and shale gas supply chain design. Comput. Chem. Eng. 92, 230–255.
U.S. EPA, 2016. Technical Development Document For Effluent Limitations Guidelines and Standars for the Oil and Gas Extraction Point Source Category. Washington, DC.
Yang, L., Grossmann, I.E., Mauter, M.S., Dilmore, R.M., 2015. Investment optimization model for freshwater acquisition and wastewater handling in shale gas production. AIChE J. 61, 1770–1782.
Zore, Ž., Čuček, L., Kravanja, Z., 2017. Syntheses of sustainable supply networks with a new composite criterion – Sustainability profit. Comput. Chem. Eng. 102, 139–155.

534130 A Multi-Objective Energy-Water Nexus Planning Model: A Case Study of the Power Systems in Texas Edwards Aquifer

Abstract:

Increasing power demands and water scarcity as well as environmental concerns give rise for the need of a planning strategy for power generating systems that takes into consideration the nexus between energy and water [1]. Power systems planning problems are complex and comprise of decisions related to: (i) the ability to add additional power plants with different types of cooling and power generating technologies, (ii) the ability to convert the cooling and or power generating technologies of existing power plants, (iii) uncertainties, and (iv) environmental and social impacts [2]. Given that the type of power generating and cooling technology of a power plant directly affects its water usage, the optimization of a power system is a multi-objective problem [3]. Therefore, a systematic methodology based on the Energy-Water Nexus (EW-N) is necessary for modeling and solving the optimization problem on power systems planning [4].

In this work, we develop a novel superstructure for planning strategies of power generating systems based on EW-N connections. The EW-N problem is formulated as a two-stage stochastic multi-objective mixed integer linear program [5]; with "here-and-now" decision variables such as additional power plants with specific power generation and cooling technologies and "wait-and-see" decision variables comprising of and not limited to the capacity and water utilization of power plants. The model determines the optimal conversion and expansion policies for the current infrastructure, based upon uncertain future water availability and power demands. The model is implemented into a case study regarding EW-N in the Edward Aquifer, aiming to provide policy makers with a systematic tool to analyze various scenarios and technology options. Moreover, the model is also flexible in that it can also be integrated into the Edwards Aquifer Groundwater and River System Simulation Model (EDSIMR) [6, 7], which is a regional model for the comprehensive decision-making and resources allocation regarding Food-Energy-Water Nexus in the Edwards Aquifer.

References:

[1] Diangelakis, N. A., & Pistikopoulos, E. N. (2017). A multi-scale energy systems engineering approach to residential combined heat and power systems. Computers & Chemical Engineering, 102, 128-138.

[2] Koltsaklis, N. E., Dagoumas, A. S., Kopanos, G. M., Pistikopoulos, E. N., & Georgiadis, M. C. (2014). A spatial multi-period long-term energy planning model: a case study of the Greek power system. Applied Energy, 115, 456-482.

[3] Mohtar, R. H., & Daher, B. (2016). Water-energy-food nexus framework for facilitating multi-stakeholder dialogue. *Water International*, *41*(5), 655-661.

[4] Garcia, D. J., & You, F. (2016). The water-energy-food nexus and process systems engineering: a new focus. *Computers & Chemical Engineering*, *91*, 49-67.

[5] Birge, J. R., & Louveaux, F. (2011). Introduction to stochastic programming. Springer Science & Business Media.

[6] Gillig, D., McCarl, B. A., & Boadu, F. (2001). An economic, hydrologic, and environmental assessment of water management alternative plans for the south central Texas region. Journal of Agricultural and Applied Economics, 33(1), 59-78.

[7] McCarl, B. A., Dillon, C. R., Keplinger, K. O., & Williams, R. L. (1999). Limiting pumping from the Edwards Aquifer: an economic investigation of proposals, water markets, and spring flow guarantees. Water Resources Research, 35(4), 1257-1268.

534167 Optimal Use of Thermal Membrane Distillation (TMD) for Treatment of Flowback Water

Abstract:

Optimal use of thermal membrane distillation (TMD) for treatment of flowback water

Kaiyu Cao[1], Priscille I. Etoughe[2], Rajib Mukherjee[2,3], Debalina Sengupta[3], Joseph S. Kwon[1,2,3*], Mahmud M. El-Halwagi[1,2,3]

1. *Artie McFerring Department of Chemical Engineering, Texas A&M Univrsity, College Station, TX*
2. *Texas A&M Energy Institute, Texas A&M University, College Station, TX*
3. *Gas and Fuels Research Center, Texas A&M Engineering Experiment Station, College Station, TX*

The flow rate of flowback water from hydraulic fracturing varies with time. Most of the water are produced within the first two weeks after hydraulic fracturing is completed. The concentration of total dissolved solids (TDS) in the flowback water also changes with time. For efficient recovery and reuse of flowback water, treatment with thermal membrane distillation (TMD) for the removal of TDS has been found to be the method of choice [1]. Even though flowback water is a continuous process, in this work, for modeling purpose a multi-period optimization problem has been formulated to account for the time-dependent variation of the flow rate and TDS concentration of flowback water, accounting for the time-; hereeach day is treated as a period. Then, we have integrated a TMD unit, considering the largest capacity of flowrate from a well in any given period. An upper limit on the concentration of TDS in the discharged (reject) water is considered to avoid precipitation. Utilizing the field data of flowback water from a given well for each period, the optimization formulation has been solved to determine the maximum period of time for optimal use of TMD, maximizing annualized profit from recovered water per period. Our result shows that it is not profitable to use water recovery beyond first few days, depending on the formation. The performance of the optimization formulation is measured in terms of percentage of recovery and is compared with cost associated with them.

References:

[1] Elsayed, N. A., Barrufet, M. A., Eljack, F. T., & El-Halwagi, M. M. (2015). Optimal design of thermal membrane distillation systems for the treatment of shale gas flowback water. *Int J Membr Sci Technol*, *2*, 1-9.

536214 The Energy-Water Nexus of Thermoelectric Power Generation and Its Impacts in the Muskingum River Watershed in Ohio

Abstract:

Thermoelectric power plants generate electricity and flue gas scrubbers mitigate emissions from the plants. However, at the same time the power plants withdraw large quantities of water from the watershed, which corresponds to 45% of the total 2010 US water use [1], and emissions from the scrubbers cause deposition of nutrients in the watershed. Therefore, increasing electricity generation could increase water stress and deteriorate water quality. Also, in the watershed where power plants are located, other activities, such as farming, interact with the power plants since these activities also require water as well as energy from the power plants and release nutrients to the watershed. To prevent shifting of environmental impacts across multiple flows [2], the Energy-Water nexus between different activities in the watershed needs to be understood in assessing the impacts of power plants.

Muskingum River Watershed (MRW) in Ohio ranks fourth as the most polluted watershed in the US [3], and two coal-fired power plants and three natural gas-fired power plants are located in the MRW. In this work, a holistic analysis approach is employed to evaluate the water-energy nexus in the MRW. Various technological and ecological systems are considered. Technological systems include mining, thermoelectric power generation, and CO_2 conversion activities. Ecological systems include land use and farming practice. Alternative scenarios for each component are analyzed to address the nexus. An agro-ecological model, which is developed by the Soil and Water Assessment Tool (SWAT), is used to assess water quality and quantity [4], while engineering models are used for the power plants and scrubbing processes.

Fossil fuel power plants not only require a huge amount of water, but also emit 28% of the 2016 US greenhouse gas emissions [5], 67% of the 2014 US SO_2 emissions, and 12% of the 2014 US NO_x emissions [6]. Therefore, ecosystem services, such as water quality regulation, air quality regulation, and climate change regulation, play an important role in mitigating nutrient runoffs and air emissions. To account for the role of ecosystems, Techno-Ecological Synergy framework is applied in analyzing the impacts of activities in the MRW [7,8]. In this analysis, the trade-offs between water quality, water quantity, net electricity generation, carbon footprint, air quality, and net present value are identified. Various scenarios are considered to suggest better solutions that are 'win-win' in terms of multiple objectives.

Considering the Energy-Water nexus in the analysis will avoid shifting of the environmental impacts across multiple flows because interactions between flows are identified and all flows are included as objectives [2]. This work could be applied to any watershed and could be extended to address the Food-Energy-Water nexus by including food productions from farming activities as another objective. Addressing the nexus could help businesses and policy makers in making decisions to improve the sustainability of the watershed.

References

[1] Maupin, M.A. et al. Estimated use of water in the United States in 2010. U.S. Geological Survey, Reston, Virginia (2014).

[2] Bakshi, B. R., Gutowski, T. G. & Sekulic, D. P. Claiming Sustainability: Requirements and Challenges. ACS Sustain. Chem. Eng. 6, 3632–3639 (2018).

[3] Inglis, J. & Dutzik, T. Wasting our waterways: Toxic industrial pollution and restoring the promise of the clean water act. Environment America Research & Policy Center, Boston, MA (2006).

[4] Khanal, S., Lal, R., Kharel, G. & Fulton, J. Identification and classification of critical soil and water conservation areas in the Muskingum River basin in Ohio. J. Soil Water Conserv. 73, 213–226 (2018).

[5] Inventory of U.S. greenhouse gas emissions and sinks: 1990-2016. U.S. Environmental Protection Agency, Washington, D.C. (2018).

[6] 2014 National Emissions Inventory (NEI). US Environmental Protection Agency. https://www.epa.gov/air-emissions-inventories. Accessed on Apr 13 2018.

[7] Bakshi, B. R., Ziv, G. & Lepech, M. D. Techno-ecological synergy: A framework for sustainable engineering. Environ. Sci. Technol. 49, 1752–1760 (2015).

[8] Gopalakrishnan, V., Bakshi, B. R. & Ziv, G. Assessing the capacity of local ecosystems to meet industrial demand for ecosystem services. AIChE J. 62, 3319-3333 (2016).

537140 Systematic Analysis and Optimization of Water-Energy Nexus

Abstract:

Securing scarce energy and water resources from the environment, while at the same time satisfying on-growing water and energy demands for end-use consumption is crucial [1]. The intermediate processing facilities, energy sources (acting also as water sinks) and water sources (acting as energy sinks), receive energy and water natural resources respectively, process, exchange and deliver them as usable electricity, fuels and water for consumption. It is important to quantify the performance and the efficiency of this water-energy nexus, and understand how it is affected by the selection of the energy and water sourcesâ€™ technologies, operating capacities and their connectivity.

To this end, we present a scalable, graph-theoretic representation of a nexus for the design and optimization of complex water-energy systems [2]. We define a nexus as a directed bipartite graph with water and energy flows and show that for specified external grid demands, the optimal nexus configuration with minimum water and energy generation is the one without any redundant subgraphs. We then introduce the water-energy (WEN) diagram that can systematically identify and eliminate those redundant subsystems. This leads to (i) minimum generation/extraction of water and energy resources from the environment, or (ii) maximum yield of water and energy to meet external demands. Our approach is simple to implement and results in optimal nexus configurations that also consider operational constraints, connectivity restrictions and water quality specifications. In the second part, we develop mathematical programming-based algebraic method that provides the same results but for more complicated and large-scale nexus design. This also takes into account the intensity factors of the sources, restrictions on source-sink matching, and quality specifications. The model is further extended to include capital and operating costs of the processing facilities, as well as geographic distances. The approach is demonstrated using case studies on water-energy nexus systems focusing on power generation, seawater desalination, groundwater and surface water at regional and national scales.

References:

[1] D. J. Garcia and F. You, "The water-energy-food nexus and process systems engineering: a new focus," Computers & Chemical Engineering, vol. 91, pp. 49-67, 2016.

[2] S. Tsolas, M. N. Karim, M. M. F. Hasan, " Optimization of Water-Energy Nexus: A Network Representation-based Graphical Approach," Applied Energy, *Submitted.*

538380 Thermo-Economic Optimization Based Comparison of Membrane Distillation Vs Mechanical Vapor Recompression for Shale Gas Produced Water Treatment

Abstract:

With advancements in horizontal drilling and hydraulic fracturing methods, unconventional shale gas extraction is rapidly developing. However, shale gas production is accompanied with a host of environmental challenges. One such critical challenge is the sustainable management of large quantities of high salinity wastewater generated in the process of hydraulic fracturing. While recycling may be a short-term strategy, long term solutions are needed for economical and environmentally conscious management of high salinity wastewaters. Desalination is one possible technological solution to treat the produced water and reduce the brine volume for final disposal while simultaneously producing fresh water for other beneficial uses. Membrane distillation (MD) system is a vapor pressure driven emerging membrane technology capable of handling hypersaline wastewater. Since MD can be operated at low temperature (30-90^{0}C), it has the potential to be integrated with low-grade waste heat sources. Another advantage of MD is the low capital investment of the process. However, low energy efficiency of MD results in high operating costs when waste heat is not available. Mechanical vapor recompression (MVR) is another desalination technology capable of treating hypersaline produced water. Unlike MD, MVR system is a well-established work driven desalination process, in which the heat recovery and energy efficiency is relatively high by utilizing the vapor compression cycle. In contrast with MD, MVR system requires considerable amount of high quality energy sources. Considering the competitive desalination efficiency of these two technologies, there is a need for a systematic way to comparison their energetic and economic performance. To date, very limited work exists on techno-economic optimization and evaluation of MD and MVR technologies for produced water treatment.

We present a mixed-integer non-linear programming (MINLP) formulation of a direct contact membrane (DCMD) process design for produced water treatment while minimizing the total treatment cost. The proposed model simultaneously optimizes the operating conditions (i.e. the temperature, pressures, and flowrate of all streams) and DCMD plant configuration (number of membrane stages, number of modules in each stage, and membranes geometry). Also, the optimization model, takes into account the possibility to improve system performance by including brine recirculation and heat integration. The results of the proposed optimization model are compared with our previous work on MVR optimization for application in produced water treatment. The optimization model is tested for a hypothetical desalination plant with 0.5 million gallons per day (MGD) capacity, in which produced water with 10% salinity is concentrated to 30% salinity of the exiting brine. The results are also validated through ASPEN simulations. Sensitivity analysis is also conducted to analyze the effect of input parameters on the system performance and economics. Tradeoffs between MD and MVR technologies for treatment of shale gas produced water will be discussed.

529522 Application of Adsorbate Solid Solution Theory to Design Novel Adsorbents for Arsenic Removal Using Computer-Aided Molecular Design

Abstract:

Application of Adsorbate Solid Solution Theory to Design Novel Adsorbents for Arsenic Removal Using CAMD

Rajat K. Doshi[1,2,], Arti Ayodhyaprasad Rajput[3], Rajib Mukherjee[1], Suresh Gupta[3] , Urmila M. Diwekar[1*],

1. Vishwamitra Research Institute, Crystal Lake, IL, USA
2. Henry B. Plant High School, 2415 S. Himes Ave., Tampa, USA
3. Department of Chemical Engineering, Birla Institute of Technology & Science, Pilani, India

*Corresponding: Urmila@vri-custom.org

Arsenic is a carcinogenic contaminant that pollutes the groundwater, a consequence of poor arsenic disposal. Various techniques are used to remove arsenic, such as oxidation, coagulation-flocculation, ion exchange, and adsorption, among which adsorption is the most efficient method. Current arsenic separating agents have a limited adsorption capacity. The overall objective of this work is to develop a computational tool for the design of novel adsorbents for arsenic remediation using clay materials including beidellite, zeolite, and sepiolite that are cheap and readily available [1]. In the first part of this research, we use the Group Contribution Method (GCM) to predict thermodynamic properties and calculate the UNIFAC interaction parameters between arsenic and other functional groups that are selected from clay materials. In the second part of this research, we utilize a computer-aided molecular design (CAMD) framework that develops new adsorbent candidates with enhanced adsorption capacities based on the group interaction parameters generated in the first part. The efficient ant colony optimization (EACO) algorithm maximizes the adsorption capacity with certain structural possibilities, thermodynamic property correlations, and process conditions. It was found that the newly designed adsorbent has an order of magnitude higher removal capacity than the adsorbents' reported in the literature [2]. The adsorption capacity estimated for the designed adsorbent was simulated in groundwater conditions; however, arsenic contamination is not solely confined to groundwater as it is also highly prevalent in wastewater flows from coal-burning power plants. An expansion of our work will be carried out by conducting experiments on the designed adsorbents. The goal is to test the performance of the designed adsorbent in coal power plant's wastewater flows and understand how the different ionic environment affects the adsorption capacity. The maximum amount of arsenic that the designed adsorbent can adsorbed per unit mass of adsorbent from wastewater found in coal-burning power plants will be presented. Moreover, the novel adsorbent has also been synthetically prototyped to determine the actual adsorption capacity of the novel adsorbent in varying ionic environments and arsenic concentrations. Our initial results from conducting a set of experimentation has demonstrated an enhanced adsorption capacity of the synthetic adsorbent that is higher than any naturally occurring adsorbent.

References:

1. Bektaş, N., Aydın, S., & Öncel, M. S. (2011). The adsorption of arsenic ions using beidellite, zeolite, and sepiolite clays: a study of kinetic, equilibrium and thermodynamics. Separation Science and Technology, 46(6), 1005-1016.
2. Doshi, R. K., Mukherjee, R., & Diwekar, U. M. (2018). Application of Adsorbate Solid Solution Theory To Design Novel Adsorbents for Arsenic Removal Using CAMD. ACS Sustainable Chemistry & Engineering, 6(2), 2603-2611.

542620 The Impact of Shale Gas and Oil on the Chemical Industry

Abstract:

Lower alkanes and aromatics are very useful as both fuels and chemical raw materials. Occasionally supply disruptions constrained availability but also encouraged investigation of alternative feedstocks and chemistries. Recent new technologies for the production of unconventional gas and oil sources from tight shale formations significantly increased availability of both methane and condensate, radically altered process economics, resulted in the construction of much new intermediates capacity, and multiplied the ways traditional and alternative chemistries might be exploited. This presentation will outline historically how natural gas components became useful as chemical feedstocks, how government regulations occasionally led to perturbations in availability and price, how supply crises led to the development of alternative chemistries, catalysts, and ultimately the exploitation of shale resources. More specifically, will be discussed how excess coproducts, especially condensate, from increasing shale gas and oil production for energy are having significant unexpected impacts on the chemical process industries and opportunities for new process chemistry research.

541186 Sustainable Energy and Chemicals: Past, Present, and Future

Abstract:

The last 40 years has seen the U.S. energy situation vacillate from gasoline rationing and concerns over Peak Oil, to projected energy and feedstock prices that were noncompetitive in global markets for chemicals production, to the economic revolution of frack gas and light tight oil. The importance of low cost energy and feedstocks in driving economic opportunity will be discussed relative to this history, along with the importance of sustainability in decision making to gain stakeholder buy-in and acceptance. Strategies for managing career paths and competencies despite changing targets and high uncertainty will be addressed.

542618 Disruptions: What the Future May Hold

Abstract:

Disruptive forces are all around us. They change the way we communicate and work. They change the way we develop goods and services. They also change the types of products that are manufactured. Examples such as the way in which we produce and consume energy or the feedstocks we use to produce product have seen many changes over the course of the last few decades, and these disruptions are expected to continue. The chemical engineering discipline and the expectations for a chemical engineerâ€™s skills has also changed and adapted as a result of these disruptive forces. A discussion of the broad themes associated with these forces and the implications that they may have to chemical engineers in the future will be discussed.

542621 Geopolitical Factors Influencing the Evolution of the Chemical Industry

Abstract:

The basic chemicals and plastics industry is built on a simple principle: find a cheap feedstock and convert it into something more valuable. However, the right feedstock depends on place and time. Certain policies and actions of various governments have influenced the evolution and profitability of the chemical industry during the past 50 years, sometimes in surprising ways. For example, high oil prices and US government funding encouraged the development and commercialization of coal gasification technology during the late 1970s and early 1980s. This enabled China to develop a competitive basic chemical industry using coal as a feedstock during the last 25 years. Similarly, an abrupt increase in oil prices during the first decade of the 21st century encouraged the present shale gas revolution, causing a major shift in the economy of producing basic chemicals and polymers in the US. If history is our guide, then what governments choose to fund and regulate may be as important as market forces in determining the shape of our industry and profession in the future.

542622 Agility & Resilience: How to Maintain Career Competitiveness in the Changing Chemical Industry

Abstract:

The world around is changing. The shift to protectionism, global instability and price volatility of raw materials and feedstocks has the potential to wreak havoc on the chemical industry. With the revolution of shale gas in North America and the sudden drop in feedstock costs, the chemical sector in the US has reasserted its historic dominance. Learn how the role of a chemical engineer can contribute to industry growth and resilience, while adapting to allow for personal agility throughout the career life cycle.

538047 Analysis of a Circular Economy: From Food Waste to Foods

Abstract:

The US-EPA estimated that 38 million tons of food waste went to landfill in 2014 and in 2010 66.5 million tons of food didnâ€™t make it to consumers. The food waste streams total to around a billion pounds of food waste, where the majority go to landfill. Publicly owned waste water treatment plants in U.S. process approximately 14 billion kilograms of dry solids per year.[1] Turning waste streams into valuable products is the goal of the circular economy.[2]

Hydrothermal carbonization (HTC) process is proposed to be a solution to the large streams of food waste and biosolids from waste water treatment plants that creates a highly valuable, multiuse product. HTC is a thermochemical process that converts wet biomass materials to a coal-like product. Under sub-critical conditions, water in the HTC reactor stays in liquid form and acts as a reaction medium to promote the breakdown and cleavage of chemical bonds in the solid biomass. The absence of liquid-to-vapor phase change of the water makes the process significantly less energy intensive compared to a process that involves water vaporization or drying. The temperature range for HTC is from 180-240°C producing solid hydrochar as the main product.

The solid product of hydrothermal carbonization, hydrochar, has multiple applications ranging from a combustion fuel source, a catalyst for anaerobic digestion that increases the purity of methane production, a soil amendment, and the base for other carbon product such as activated carbon.

In this study we will explore the circular economy of utilizing organic wastes into foods. The steps involved in the circular economy system include: 1) collecting organic wastes (specifically food wastes and bio-solids), 2) converting the organic waste streams to hydrochar via HTC process, 3) growing food produces by indoor gardening, and 4) using hydrochar from step 2) as fuels to power the indoor gardens.

Utilizing organic waste streams in cities such as Chicago or Philadelphia, where there are a multitude of abandoned warehouses, provides an opportunity to produce locally grown produces that would typically come from across the country from California or Mexico. Not only will this circular economy reduce environmental impacts, but it will increase positive societal impacts by benefitting the local economy, providing educational opportunities of farming that would not typically be available in cities, and serving fresher produces to the local community.

A comparative life cycle assessment of our proposed process will be applied to compare conventional farming to our circular economy proposal. We hope to prove a reduction of environmental impacts, such as global warming potential, fossil fuel depletion, natural resource depletion. The life cycle assessment results will aid in a full system analysis which evaluate the proposed circular economy idea from societal, technical, environmental, economic, and political perspectives.

- https://www.epa.gov/sustainable-management-food/sustainable-management-food-basics
- https://www.sciencedirect.com/science/article/pii/S0301479717303808

*Contact: Dr. Justinus Satrio (justinus.satrio@villanova.edu)

538097 Re-Wiring the Domestic Food Trade for Reducing Irrigation Impacts in the United States

Abstract:

The food-energy-water(FEW) nexus concept recognizes the interconnected nature of the three systems and advocates managing them together to avoid sub-optimal outcomes and unintended consequences. The FEW nexus challenges differ for developed agro-economies such as the United States(U.S.) from agriculture centric developing nations. For U.S., irrigation provides a crucial link to study interconnections between food-water-energy resources. Additionally, food trade is an integral part of food system as it bridges the growing disconnect between production and consumption centers. Depending on the origin of food production, trade can potentially reduce environmental impact of food consumption. However, environmental impacts of agriculture are not directly considered when forming trade connections and may carry risk associated with suboptimal FEW system management (e.g. sourcing food from water-depleted regions). Structural optimization of food trade provides an opportunity to assess whether FEW resources can be managed together with respect to land-water-energy constraints. This work focuses on environmental impacts associated with FEW systems, namely- water required for irrigation, greenhouse gas(GHG) emissions from commodity transport, and pumping irrigation water for food production.

We leverage publicly available datasets from agriculture census and transportation surveys to create a network of regional food trade. The study combines data on state-wide food production, irrigation statistics, and life cycle inventory databases to convert food transfer network into embedded irrigation water, and GHG emission flows. Specifically, we focus on rice production and trade in the U.S. Rice is one of the staple food crops and produced entirely through irrigation, making it an ideal crop for the irrigation impact analysis. Majority of rice production is concentrated in four regions-Arkansas Grand Prairie, Mississippi Delta, Gulf Coast, and Sacramento Valley. The model optimizes rice trade network by keeping the demand constant and evaluate a series of constraints pertaining to reduction in energy and water usage to understand whether production can be shifted realistically to optimize all three systems. We limit the potential crop shifting to current production regions to account for soil and climatic parameters conducive to crop production. Additionally, we consider type of rice grown in each region as crop physiological requirements may differ based on the rice type. Preliminary results of the optimization framework reveals a rewired domestic rice trade network with a 27% reduction in virtual water use (15.1 billion m^3 reduced to 11.1 billion m^3). However, due to shift in production to more water-efficient states, a 10% increase in marginal land use is required. The land results are significant as they demonstrate the resulting sub-optimal solutions when interdependencies are overlooked. On the other hand, minimizing groundwater usage from water stressed areas also reduces overall GHG emissions demonstrating that these are not mutually exclusive goals in many states. The Pareto optimal solutions showing water-GHG emissions tradeoffs resulting from the model will be presented and their implications for regional FEW nexus sustainability will be discussed.

527168 Food, Energy, Fuels and Chemical Feedstocks from Rice Crops: Multi-Objective Optimisation of Multi-Product Value Chains for the Philippines

Abstract:

Rice is an important staple food in the Philippines such that its 20 M metric tons annual production places the country rank 8th among the rice producers in the world. However, there are many improvements in the efficiency and sustainability of the Philippines rice value chains that could be made by addressing various constraints related to technological, environmental, economic and socio-political factors [1]. Consequently, the Philippine government is striving to achieve Philippine rice competitiveness and to transform its value chains to be more profitable, resilient and sustainable; they recognise the importance of research and development as basis for such a laudable goal [2]. From a recent analysis of the Philippine rice value chain commissioned by the government, by-products like rice straw, rice hull and rice bran can be important sources of raw material for other industry sectors. Ninety-five percent of the rice bran is being used as animal feed component while seventy percent of the of the rice hull is being used as an alternative source of energy and fuel. Moreover, the silica-rich rice hull ash is being used as additive in the cement industry [3]. These could not only provide affordable renewable energy, fuels and materials but also much-needed rural development and employment opportunities. However, there are complex issues that need to be addressed: rice production is a water- and energy-intensive activity. The global water footprint from rice production is 1308 Mm^3/year with irrigated rice cultivation consuming the most water. Energy is required for the production of fertilisers, electricity for pumping water and heat for parboiling, drying and other milling activities [4]. Furthermore, the limited land for rice cultivation is being threatened by the rapid and unregulated conversion of agricultural land to other uses [1]. In this project, a comprehensive optimisation model, based on mixed-integer programming [5-6], is being developed that can support complex decision-making related to multi-product rice value chains and can capture the trade-offs between water, energy and land utilisation. The model is used to identify promising rice value chains for sustainable and efficient production of food, energy, fuels and chemical feedstocks. In this conference, we will present the optimisation model and the key findings from the case studies.

References:

[1] B.A. Excode (2016). Rice production and consumption trends in the Philippines from 2000 to 2015. Yuchengco Center, De La University; Philippines.

[2] C.T. Briones (Ed.) (2016). The PhilRice strategic plan 2017-2022. Philippine Rice Research Institute; Philippines.

[3] A.B. Mataia, A.B. Mataia, R.G. Manalili, J.C. Beltran, B.M. Catudan, N.M. Francisco and A.C. Flores (2016). Analysis of the rice value chain in the Philippines. Philippine Rice Research Institute; Philippines

[4] R. Sims, A. Flammini, M. Puri, S. Bracco (2015). Opportunities for Agri-Food Chains to become energy-smart. Food and Agriculture Organization of the United Nations and the United States Agency for International Development; United States of America.

[5] S. Samsatli, N.J. Samsatli (2018). A multi-objective MILP model for the design and operation of future integrated multi-vector energy networks capturing detailed spatio-temporal dependencies. Applied Energy. DOI: 10.1016/j.apenergy.2017.09.055.

[6] S. Samsatli, N.J. Samsatli, N. Shah (2015). BVCM: a comprehensive and flexible toolkit for whole-system biomass value chain analysis and optimisation - mathematical formulation. Applied Energy, 147, pp. 131-160.

541194 Modeling the Impacts of International Food Trade on Contaminant Transport and Human Exposure

Abstract:

Introduction: Quantifying and addressing the implications of the transfer of resources across borders is a hallmark of research on the food, energy and water (FEW) nexus. International trade in goods can result in severe imbalances in the water stress, green house gas emissions, and energy consumption experienced by or attributed to producing and receiving countries[1,2]. Yet an important component of food trade—that is, the potentially hazardous chemicals contained in foods as a result of accumulation from production environments, processing, transport, or storage—is still poorly quantified[3–5]. Here, we investigate the importance of international food trade on the dietary exposure of the Swiss population to polybrominated diphenyl eithers (PBDEs). PBDEs are lipid soluble compounds used as flame retardants in synthetic fibers and molded plastics. Of major concern is PBDE exposure through the diet. Studies have confirmed that fish, meat, and dairy products contribute most to the daily PBDE intake[6]. Like other countries, Switzerland has reported a gradual shift in dietary patterns over the past few decades, with increasing consumption of fish reported, reaching a peak value of 17.6 kg in 2010, and with a shift in the origin of fish. We therefore constructed a model to quantify the exposure of the Swiss population to PBDEs from local and imported fish and seafood.

Methodology:

Model boundaries: Our study focused on Switzerland as a geographical case study. All calculations referred to fish imports for the year 2016. Fish, shrimp/prawn, and mussels were selected as key fish and seafood types to include in the study, as they are among the most highly consumed.

Data collection: An extensive literature review was conducted to derive useful input data for our model, including PBDE concentrations in fish, shrimp and mussels from different countries, and to estimate daily per capita fish consumption in Switzerland. Trade data (imports, exports, and re-exports) for fish, crustaceans (shrimp and prawns) and mollusks (mussels) was extracted from the UN Comtrade Database. Data were collected for fresh, frozen, fresh fillet, and frozen fillet categories of the respective commodities.

Model Structure: The model consists of four primary components. These are: (i) PBDE concentrations in fish from different origins, resulting from bioaccumulation from their environment and/or food; (ii) food trade data for Switzerland; (iii) local fish production; and (iv) dietary intake to the Swiss population via fish and seafood consumption.

Results and Discussion: Based on the difference between average per capita fish consumption, production, imports and re-exports, we estimate that Switzerland imports approximately 66% of fish consumed, while only 34% is locally caught. Of the imported fish, Norway, Denmark and the UK are the largest exporters of salmon to Switzerland. For shrimp and prawn, the majority is from Vietnam. For other fish, Norway, Denmark, the Netherlands, Vietnam, France, Germany, Italy, and China are the largest exporters. Within the locally produced fish, whitefish (*Coregonus sp.)* is the most common fish contributing to 60% of local the catch, followed by perch (19%), pike (7%) and roach (7%)[7]. Using an average of 23g/day fish consumption, results show that from 70000 metric tons fish consumed in Switzerland in 2016, 66 % (46000metric tons) was imported corresponding to 34 % locally caught fish (24000 metric tons).

Based on published studies of PBDEs in fish, the highest concentrations were detected in whitefish (Coregonus spp.) in Switzerland[8]. Norway also reported high levels of PBDEs in whitefish[9]. The US reported PBDEs as high as 3078 pg/g wet weight in salmon[10]. Compilation of country-specific PBDE body burdens in fish continues. In addition, we are employing modeling approaches to estimate PBDE body burdens in key species from exporting countries where data are lacking. These data will aid in forming a comprehensive source-receptor relationship for Swiss population exposure to PBDEs via seafood and fish.

References Cited

(1) Vora, N.; Shah, A.; Bilec, M. M.; Khanna, V. Food–Energy–Water Nexus: Quantifying Embodied Energy and GHG Emissions from Irrigation through Virtual Water Transfers in Food Trade. *ACS Sustain. Chem. Eng.* **2017**, *5* (3), 2119–2128.

(2) Leck, H.; Conway, D.; Bradshaw, M.; Rees, J. Tracing the Water–Energy–Food Nexus: Description, Theory and Practice. *Geogr. Compass* **2015**, *9* (8), 445–460.

(3) Ng, C. A.; von Goetz, N. The Global Food System as a Transport Pathway for Hazardous Chemicals: The Missing Link between Emissions and Exposure. *Environ. Health Perspect.* **2017**, *125* (1), 1–7.

(4) D'Odorico, P.; Carr, J. A.; Laio, F.; Ridolfi, L.; Vandoni, S. Feeding Humanity through Global Food Trade. *Earths Future* **2014**, *2* (9), 2014EF000250.

(5) Ercsey-Ravasz, M.; Toroczkai, Z.; Lakner, Z.; Baranyi, J. Complexity of the International Agro-Food Trade Network and Its Impact on Food Safety. *PLoS ONE* **2012**, *7* (5), 1–7.

(6) Schecter, A.; Haffner, D.; Colacino, J.; Patel, K.; Päpke, O.; Opel, M.; Birnbaum, L. Polybrominated Diphenyl Ethers (PBDEs) and Hexabromocyclodecane (HBCD) in Composite U.S. Food Samples. *Environ. Health Perspect.* **2010**, *118* (3), 357–362.

(7) Swiss fact: around 1.8 million kilograms of fish are caught in Switzerland every year https://lenews.ch/2017/01/04/swiss-fact-around-1-8-million-kilograms-of-fish-are-caught-in-switzerland-every-year/ (accessed May 31, 2018).

(8) Zennegg, M.; Kohler, M.; Gerecke, A. C.; Schmid, P. Polybrominated Diphenyl Ethers in Whitefish from Swiss Lakes and Farmed Rainbow Trout. *Chemosphere* **2003**, *51* (7), 545–553.

(9) Mariussen, E.; Fjeld, E.; Breivik, K.; Steinnes, E.; Borgen, A.; Kjellberg, G.; Schlabach, M. Elevated Levels of Polybrominated Diphenyl Ethers (PBDEs) in Fish from Lake Mjøsa, Norway. *Sci. Total Environ.* **2008**, *390* (1), 132–141.

(10) Schecter, A.; Päpke, O.; Tung, K.-C.; Staskal, D.; Birnbaum, L. Polybrominated Diphenyl Ethers Contamination of United States Food. *Environ. Sci. Technol.* **2004**, *38* (20), 5306–5311.

523949 Using Agricultural Wastes to Recover Rare Earth Elements from End-of-Life Materials

Abstract:

An economical and eco-friendly bench-scale process that relies on microbial production of organic acids has been developed to solubilize and release rare earth elements (REE) from solid recyclable materials (Reed, Hydrometallurgy, 2016, 166:34-40). Techno-economic analysis (TEA) and life cycle analysis (LCA) for bioleaching of REE from fluid catalytic cracking (FCC) catalysts suggested that a bioleaching plant could be profitable, but could be even more profitable if the cost of the sugar (glucose) used for growing the microbes was reduced (Thompson, ACS Sustainable Chemistry & Engineering, 2017, 6:1602-1609). This presentation will focus on our efforts to improve the economics of REE bioleaching using alternative carbon and energy sources as microbial growth substrates.

In our previous work examining the potential for several organic acid producing microorganisms to leach REE from end-of-life waste materials we identified *Gluconobacter oxydans* B58 as a promising industrial strain. Lixiviant produced by *G. oxydans,* containing predominantly gluconic acid, could recover ~50% of the total REE content of FCC catalyst. The biologically produced lixiviant was more effective than an abiotically prepared solution containing a similar concentration of gluconic acid. TEA and LCA showed that bioleaching could be more profitable and environmentally friendly than industrial standard chemical leaching processes, suggesting commercial viability. However, substantial reductions in operating costs could be achieved by replacing the refined glucose used in the original growth medium with a lower cost substrate. Consequently corn stover and potato wastewater were tested as growth substrates for *G. oxydans*. The initial lixiviants generated were tested against FCC and yielded a leaching efficiency 50-90% of the best case leaching efficiency of glucose-derived biolixiviant. Work is ongoing to optimize biolixiviant production from agricultural wastes. The conditions for improving bioleaching using waste-derived biolixiviants, including TEA and LCA evaluations, will be discussed.

526360 Membranes for Nutrient Concentration, Industrial Separations and Applications Beyond

Abstract:

With the increasing importance of the waterâ€"energyâ€"food nexus, it is critical that membranes can be designed for optimal performance and energy efficiency for large varieties of water and food applications. The cost and energy consumption of milk and whey concentration systems can be minimized when one selects a reverse osmosis element that reduces fluid flow requirements while maintaining optimum dewatering efficiency. In addition, superior performance of RO membranes which address the emerging importance of high rejection will be discussed for nutrient concentration, with a properly designed membrane based on the understanding of membrane structure-property relationships. Potentially there are many more applications where RO technology can be used to meet critical customer requirements with high efficiency and favorable economics.

AUTHOR INDEX